Helicopter Pilots

Annette Smith

Contents

Helicopters

Dana, Rick, Jake and Ali are helicopter pilots.
They work at an airport.

There are lots of helicopters at this airport.
Some of them are very big.
They are rescue helicopters.
Some of them are much smaller.
These helicopters are for short trips.

Dana flies the smaller helicopters.
Rick and Ali fly the bigger ones.
Jake has been trained to fly helicopters for fighting fires.

Dana flies people to beautiful places in her helicopter.

Dana

Flying Over Land and Water

Dana takes people on helicopter flights over mountains, forests and lakes. As she flies over these places, she talks to the people about the animals and birds that live there.

Many people want to take photos.
Dana flies the helicopter down
closer to the trees and water.
She is always careful to watch the **instruments**
in front of her so that everyone is safe.

Rescuing People

Rick

Rick is always ready
to help rescue people
if they are in danger.
He often gets a phone call
to fly out to a beach.
Sometimes, a swimmer
needs to be rescued
because enormous waves
are pushing them way out to sea.

Rick and his **co-pilot** work together very carefully
to fly the helicopter down close to the water.
A **lifesaver** goes down on a rope
to get the swimmer.
Then, they are pulled up together
into the helicopter.

A lifesaver and a swimmer are pulled up into the helicopter.

People who climb mountains sometimes slip down cliffs and hurt themselves.

Rick gets a phone call and flies out to help them. He takes a doctor as well as a co-pilot. The doctor and the co-pilot rescue the climber with a ladder.

A doctor helps a climber before they are lifted up to the helicopter.

Then, Rick flies the helicopter to a hospital. Many big hospitals have a **helipad** on the roof. Rick lands the helicopter on a helipad on the hospital roof.

Putting Out Fires

In hot, dry weather,
fires often move very fast
across the land.
Trees and long grass
burn quickly.
Clouds of smoke fill the air.
People, animals and their
homes are in danger.

Jake

Jake flies one of the big helicopters over the fire.
There are tanks full of water in the helicopter.
Jake lets the water pour out over the fire.

Jake drops water from his helicopter onto a fire.

Taking People to Hospital

Helicopters are needed
at some road accidents.
If the roads are busy
with cars and trucks,
an ambulance cannot get
through the **traffic**.
So the police make a place
for Ali to land her helicopter.

Ali

The doctor in the helicopter helps the police at the accident.
They put the people who are hurt into the helicopter.
Then, Ali flies them quickly to a hospital.

Helping People

Today, Ali is on her way out to the country.
A big boy has fallen off his mountain bike.
He has broken his arm badly.
His parents want to get him to the hospital quickly.
Ali finds the mountain bikers
and lands the helicopter.

Soon, they are on the way to the hospital.

Dana, Rick, Jake and Ali love their work as helicopter pilots.

They are very happy that they can help people who need them.

Glossary

co-pilot (*noun*) someone who helps the pilot fly a helicopter

helipad (*noun*) a landing place for helicopters

instruments (*noun*) tools for checking how high or how fast a helicopter is going

lifesaver (*noun*) someone trained to rescue people from the water and give first aid

traffic (*noun*) cars and trucks moving along a road